Thomas Henry

An Account of a Method of Preserving Water,

at sea, from putrefaction, and of restoring to the water its original

pleasantness and purity, by a cheap and easy process: to which is added a

mode of impregnating water, in large quantities

Thomas Henry

An Account of a Method of Preserving Water,

*at sea, from putrefaction, and of restoring to the water its original pleasantness
and purity, by a cheap and easy process: to which is added a mode of impregnating
water, in large quantities*

ISBN/EAN: 9783337393267

Printed in Europe, USA, Canada, Australia, Japan

Cover: Foto ©berggeist007 / pixelio.de

More available books at **www.hansebooks.com**

A N

ACCOUNT OF A METHOD OF

PRESERVING WATER, AT SEA,

F R O M

P U T R E F A C T I O N,

A N D O F

RESTORING TO THE WATER

ITS ORIGINAL PLEASANTNESS AND PURITY,

BY A CHEAP AND EASY PROCESS:

TO WHICH IS ADDED

A MODE OF IMPREGNATING WATER,

IN LARGE QUANTITIES,

W I T H F I X E D A I R,

For MEDICINAL USES, on BOARD SHIPS, and in HOSPITALS;

AND LIKEWISE

A PROCESS FOR THE PREPARATION OF

ARTIFICIAL YEAST.

By THOMAS HENRY, F. R. S.
And MEMBER of the MEDICAL SOCIETY of LONDON.

W A R R I N G T O N:
PRINTED BY W. EYRES FOR J. JOHNSON, N°. 72,
ST. PAUL'S CHURCH YARD, LONDON.
MDCCLXXXI.

TO THE RIGHT HONOURABLE

J   O   H   N

EARL OF SANDWICH, &c.

FIRST LORD COMMISSIONER,

AND TO THE OTHER

LORDS COMMISSIONERS

FOR EXECUTING THE OFFICE OF

LORD HIGH ADMIRAL OF GREAT BRITAIN,

THE FOLLOWING ACCOUNT OF A

METHOD OF PRESERVING WATER, AT SEA,

FROM PUTREFACTION, &C.

I S,

WITH THEIR LORDSHIPS' PERMISSION,

MOST RESPECTFULLY INSCRIBED

BY THEIR DEVOTED AND OBEDIENT

HUMBLE SERVANT,

MANCHESTER,
May 24, 1781.   THOMAS HENRY.

PREFACE.

THE Author of the following Account is by no means ignorant of the advantages which may be derived from the mode of diftilling Sea-Water, which has been lately invented ; but he apprehends it is intended only to furnifh a partial fupply of frefh water, in cafes of neceffity, and infufficient for the confumption of a large crew : whereas his fcheme aims at the conftant prefervation of water, from the putrefaction to which it is expofed in long voyages. The fame procefs, alfo, which is directed for the precipitation of the lime, may, with a

fmaller

fmaller portion of the effervefcing fub-
ftances, be commodioufly ufed for the
fpeedy reftoration of the air of which
water is deprived by diftillation ; and
thus the objections which have been
made to the want of fpirit and fuppofed
infalubrity of diftilled water will be ef-
fectually removed, whenever the confumption of a fhip's ftore of water fhall
render it neceffary to have recourfe to
diftillation.

E R R A T U M.

Page 26, line 19, *for* remained *read* remains.

AN ACCOUNT OF A METHOD OF

PRESERVING WATER

O N

SHIP-BOARD

F R O M

PUTREFACTION, &c.

THAT the ftrength, fafety and wealth of this nation depend principally on her Marine, is a fact, the truth of which has been long eftablifhed. The prefervation, therefore, of the health and lives of our Seamen fhould always have been confidered as a primary object of attention, by Government.

Yet notwithftanding the ravages made by that deftructive difeafe, the fea-fcurvy, among

B   this

this valuable body of men, fo little had been done towards the prevention of the evil, that even at fo late a period as that of the late Lord Anfon's circumnavigation, the Centurion, the Commodore's fhip, before fhe had reached the ifland of Juan Fernandes, had buried two hundred of her crew, and could muſter no greater a number than fix of the common failors in a watch, who were capable of doing duty. The other two fhips which arrived at that ifland, had fuffered in proportion.

The remaining part of the crews of the three fhips, having, by means of frefh air and vegetable diet, during their continuance on the ifland, recovered from the fcurvy, em-barked on board the Centurion; but before the fhip arrived at Tinian the ficknefs re-turned, and, at the expiration of the fecond year, fo great had been the mortality, that, horrid to relate! the hiftorian of the voyage informs us, more than four fifths of the original number of thefe brave, thefe ufeful members of fociety had fallen victims to the fcurvy or putrid fevers.

How

How great the contraſt between this me-
lancholy deſcription, and the pleaſing narra-
tive which the ever to be lamented captain
Cook has given of a voyage equally long in
its duration, and more variable in climate.
In this voyage, in which he paſſed through
all the climates from 52 deg. N. to 71 S.
with a company of one hundred and eighteen
men, and the performance of which employed
no leſs a period than three years and eighteen
days, this prudent, this humane Commander
loſt only one man by diſeaſe; and it even
appears that this man carried on board with
him the primary ſymptoms of the diſorder
which, afterwards, deſtroyed him.

Much praiſe is due to Government for the
many ſalutary arrangements made on board
the veſſels ſent on that expedition, previous
to their ſailing; and great was the merit of
the excellent Commander of the Reſolution,
in the unremitting attention paid by him to
the uſe of every thing that could contribute,
and the avoiding of every thing injurious, to
the health and preſervation of his men. Juſtly
did the Royal Society decree to him the

honourable reward of their annual medal, and moſt juſtly are his merits celebrated in the eloquent oration, delivered on that occaſion, by the late learned and worthy preſident, Sir John Pringle.

In the voyage which captain Cook performed, with ſuch extraordinary ſucceſs, he was particularly fortunate in meeting with frequent ſupplies of freſh water.* The iſlands in the South Sea, at which he ſtopped, abounded with ſprings of excellent water; and when, in the high ſouthern latitudes, after a long abſence from land, he was in danger of a ſcarcity of that article, the mountains of ice unexpectedly furniſhed him with an ample quantity of it; the ſalt having been precipitated from the ſea water by the act of freezing — a circumſtance which though it had been, in general, either unattended to, or

---

* "I never failed to take in water wherever it was to be procured, even when we did not ſeem to want it; becauſe I look upon freſh water from the ſhore to be much more wholeſome than that which has been kept ſome time on board."

Captain Cock's Method for preſerving the Health of Seamen. *Phil. Tranſ.* vol. LXVI. p. 405.

contro-

controverted, has been amply confirmed by experiment.*

If this had not been the cafe, and had the crew of the Refolution been under the neceffity of drinking water in that putrid ftate to which it is too frequently reduced in long voyages, it is to be feared that all the other precautions would have fallen fhort of the fuccefs that attended them. The fea-fcurvy is allowed on all hands, to be a putrid difeafe, and to be the confequence of putrid air, putrid provifions, putrid water, and the deficiences of vegetable diet and cleanlinefs.

The drinking of putrid water is not only highly difagreeable and difgufting, but extremely noxious to the conftitution. Every kind of putrid matter received into the body acts as a ferment, and fpreads its influence through the animal fyftem. Were the former the only inconvenience experienced by the gallant defenders of our ifland, the gratitude

* Mr Nairne's Experiments on Water obtained from the melted ice of fea water.

we owe them, fhould make us anxious to remove it : but when we fee thefe brave men languifhing from the want of frefh and wholefome water;* when we fee difeafes undermining thofe conftitutions which nature had formed to laft to a good old age; when we behold thofe lives, fo valuable to their country, prematurely deftroyed by the fame deleterious caufe, what zeal can be too ardent, what endeavours too active to remove the evil !

A method of preferving water free from putrefaction was fome years fince propofed by the late Dr. Alfton. It confifted in adding a quantity of lime to every cafk of water. That fubftance is known to have a ftrong antifeptic property; and water, as long as it retains the impregnation of lime, never putrifies. But the lime communicates a difagreeable tafte to the water, and, abftracted from that

---

* " I am convinced, that with plenty of frefh water, and a clofe attention to cleanlinefs, a fhip's company will feldom be much afflicted with the fcurvy, though they fhould not be provided with any of the antifcorbutics before-mentioned."

Phil. Tranf. vol. LXVI. p. 403.

incon-

inconvenience, might, perhaps, in many inftances, be detrimental.

To free the water, at the time of ufing it,
from the lime, Dr. Alfton propofed the precipitation of the latter, by throwing into the
water a quantity of magnefia alba; on this
principle; that as lime-ftone is rendered foluble in water by its deprivation of fixed air,
and has a greater affinity with that air than
magnefia has, the particles of quick-lime
diffolved in the water would attract the air
from the magnefia, and thereby becoming no
longer foluble, would fall to the bottom, and
leave the water taftelefs and fit for œconomi-
cal ufes.

Dr. Alfton's theory was juft; but the expence attending it, owing to the price of
magnefia was fuch (though the expence fhould
be great indeed which is to be placed in the
balance againft the prefervation of our mari-
ners) as to prevent the execution of the
propofal.

My

My attention was directed fome time fince to this fubject, by the trouble I had often experienced from the water in the tub, intended to cool the worm of my ftill, becoming putrid, if not frequently emptied; and the labour of filling it with frefh water was irkfome to my fervant from the diftance he had to carry it. The putrefaction of it was accelerated by the neceffity of the tub being placed within a very fmall diftance of a continual fire which is kept in my laboratory, and the fetor of the putrid water was exceedingly offenfive. It occurred to me that the addition of lime to the frefh water might obviate the inconvenience. The event has far exceeded my moft fanguine expectations; for the water, though very repeatedly expofed to the heat raifed in diftillation, added to the warm fituation in which it was placed, continued free from the leaft degree of putrefaction above eighteen months, and was then thrown away, only becaufe it was become foul from duft. The water, wafted by evaporation, was fupplied from time to time by frefh quantities, but the original water was never removed during that
period,

period,* and it was conftantly expofed to the
action of the atmofpherical air.

So ftrong a proof of the antifeptic influence
of lime on water impreffed me forcibly. Dr.
Alfton's fcheme recurred to my mind, and it
appeared very defirable to invent fome method of rendering it more applicable to the
purpofes of the navy. This would be ac-
complifhed, if a cheap and eafily practicable
method could be difcovered of precipitating
the lime, and thereby reftoring the water to
its original tafte.

In order to produce this effect, I have made
a variety of experiments, with the relation of
which I fhall not detain the reader; but fhall
proceed to defcribe the method of impregnating the water with lime, and alfo the
means of precipitating the lime from the water, at the time of its ufe, and reftoring its
original fweetnefs and pleafantnefs; after which
I fhall give an account of an eafy method of
impregnating water, *in large quantities,* with

* This method would prevent much labour in large diftilleries.

fixed

fixed air, for the ufe of the fick on board his Majefty's and other fhips, as well as for hofpital practice.

The Method of impregnating the Water with Quick-Lime.

To every cafk of water of 120 gallons, add * two pounds of well burnt quick-lime, either frefh from the kiln, or preferved, as directed in a fubfequent part of this Treatife. When the lime has been in the cafk fome minutes, and the heat and effervefcence occafioned by the mixture are over, let the cafk be carefully ftopped from any communication with the external air.

* In the directions lately publifhed, by order of the French government, for preferving the health of their Seamen, which are inferted in the third number of the Medical Commentaries for 1780, they have adopted the practice of adding quick-lime, in the proportion of one pound to every whole cafk, and half a pound to each half cafk of water. Had they been acquainted with a method of again freeing the water from the lime, they would, doubtlefs, have prefcribed a larger proportion of the latter; for they acknowledge that quantity to be infufficient to preferve the water free from putrefaction : to correct which, they direct a quantity of vinegar to be added at the time the water is drawn off for ufe.

The

Fig. 3.

Fig. 1.

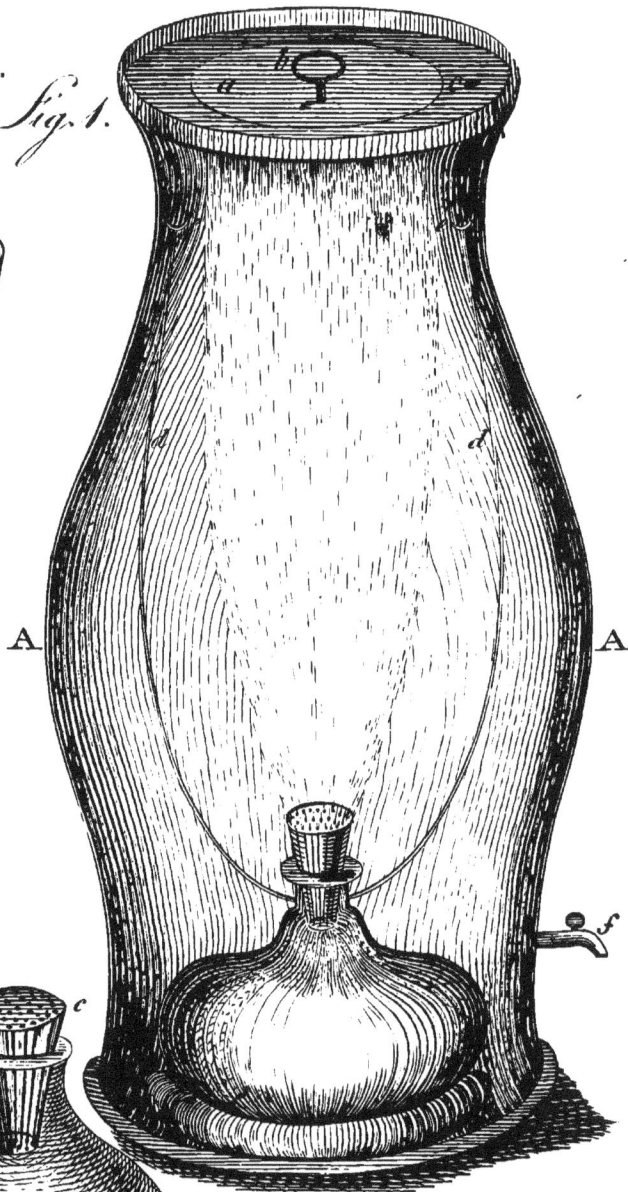

A   A

Fig. 2.

D   D

Fig. 4.

H. Clark del.   Bottomley sc.

*The Method of freeing the Water from the Lime,
and restoring it to its former State.*

Let a cask, (A A, fig. 1.) be prepared, of a form something narrower in its diameter in proportion to its depth than usual. The top, *(a)* must be formed of one plank, and have a piece cut out of the centre, of a circular form, and as large as can be allowed without weakening the sides too much.* This piece, or bung, must be made to fit as closely as possible, and have an iron handle *(b)* affixed to it, for the purpose of lifting it, and of confining a weight (fig. 4.) which is to be laid on, to keep the bung from yielding to a small force from within. A small hole *(c)* must be bored in the side of the top, which is to be exactly stopped with a plug, for a purpose to be explained in the sequel.

---

* If these sides be made of strong timber, closely cemented, they may be sufficiently firm; and the bung may be formed of a thicker plank, and turned in a lathe so as to fit very closely.

Fill

Fill this cafk, which may be fuppofed to contain fixty gallons, fecured on a convenient part of the deck, or flung up in the fhrouds, with the lime-water, drawn off clear from the fediment, fo as to avoid any vifible particles of lime floating in it; allowing fufficient room for the air veffel, and a free fpace of about half an inch between the furface of the water and the top of the cafk.

Let a veffel (D D, fig. 2.) be alfo prepared, capable of containing two gallons, or $\frac{1}{30}$ of the capacity of the cafk (A A, fig. 1.) Into this veffel introduce half a pound of marble, pure *unburnt* lime-ftone or chalk * grofsly powdered, and two quarts of water. Then pour gradually on thefe, three ounces of ftrong vitriolic acid, commonly called oil of vitriol, and ftopping the mouth of the veffel (DD) with the tubulated ftopper *(cc)* let it down by means of the ftrings *(dd)* into the cafk †

---

* I fhall, in future, make ufe of the term *mild calcareous earth* as including all thefe fpecies of it; and of thefe the preference is to be given in the order in which they are placed in the text.

† Thefe ftrings are to be faftened to the pegs *(ee)* when the air veffel is let down.

which

which is drawn tranfparent, to fhew the ftate of the whole apparatus. The fixed air, let loofe from the mild calcareous earth, will bubble up through the lime-water. When this has continued about a minute, the bung (*a*) is to be faftened on, and the weight (C, fig. 4.) flipped over the ring of the handle (*b*) to keep the bung in its place. In about an hour the bung may be removed, in order to fee whether the difcharge of air continues. If it have ceafed, or be confiderably abated, three ounces more of vitriolic acid is to be added, and the air veffel returned to its former ftation in the cafk.

The time neceffary to precipitate the lime from the water will be in proportion to the brifknefs of the effervefcence, but in general a few hours will be fufficient. Should the firft parcel of calcareous earth and vitriolic acid be unequal to the fweetening of the lime-water, and no longer difcharge air brifkly when agitated; the contents of the air veffel are to be poured out, and a frefh quantity of the ingredients fubftituted in their place.

When

When the water is become quite mild, the air veffel is to be taken out, and, if the calcareous earth continue to difcharge air, be plunged into another cafk of lime-water, that there may be no needlefs expence of the fixed air.

The fpecific gravity of the lime is fo much fuperior to that of the water, that it will foon fall to the bottom of the cafk when the operation is finifhed. As foon as the water is become clear, it muft be drawn off by the cock *(f)* for ufe; or, if the cafk be wanted to purify other quantities of water, it may be drawn off fooner into other veffels to clarify.

The precipitated lime may be collected and dried, and, being now in the ftate of chalk, and impalpably powdered, may be ufed inftead of prepared chalk for the medicinal purpofes to which that article is applied.

Cautions to be obferved in the above Proceffes.

1. The quick-lime fhould be chofen as pure, free from any foreign tafte, white, well burnt

burnt and frefh from the kiln as can be obtained. What is carried to fea for future ufe, fhould be carefully packed up in clean tight cafks, fo as to preferve it from moifture and the action of the air.

2. The cafks, into which the lime-water is put, fhould be perfectly clean and fweet; and thofe fhould be felected for this ufe that are well feafoned and free from fap.

3. The water is to be firft poured into the air veffel (DD;) then the calcareous earth, which is to pafs through a paper cone to prevent its adhering to the fides of the mouth of the veffel; and laftly, the acid is to be added, no attention being paid to the mixing the earth and water intimately. By this means the acid attacks the calcareous earth gradually, and the veffel is in no danger of burfting by the too fudden explofion of the air. For the fame reafon, care fhould alfo be taken that the air veffel be not fhaken too rapidly.

4. Gently

4. Gently agitating the upper part of the cafk from time to time, during the procefs, will accelerate the completion of it, by occafioning a quicker abforption of the fixed air. And the fmall plug muft occafionally be taken from the orifice *(c)* to let out the part of the air which is not foluble in water.

5. The precipitated lime is to be cleared out of the cafk (A A) after each time of ufing it ; and the cafk fhould be frequently wafhed out thoroughly.

6. Care muft be taken that the mouth of the air veffel be clear of calcareous earth before the ftopper be put in ; and that the ends of the tubes in the ftopper be not clogged up with any thing that may prevent the paffage of the air through them.

7. Each fhip fhould be provided with feveral of the air veffels, and each veffel fhould have two or three tubulated ftoppers adapted to it. The veffels and their ftoppers to be marked with fimilar numbers.

8. The

8. The fize and number of the purifying
cafks muft be proportioned to the rate of the
fhip, and the convenience with which they
can be managed.

9. If the cafk be left with the air veffel in
it during the night, or for any confiderable
length of time, a fmaller plug may be put into the fmall hole *(c)* in the top of the cafk,
fo as to leave it not quite air tight.

10. If during the procefs, the fixed air
fhould efcape, by the edges of the round bung,
it may be prevented by any flight luting,
which may be eafily removed when the bung
is to be taken out.*

11. It will be fcarcely neceffary to men-
tion that the air veffel and the large circular
bung-hole, in the top of the cafk, are to be
fo proportioned, that the latter will eafily
admit of the paffage of the former through
its aperture.

---

* The efcape of air will prolong, but not wholly prevent the
fuccefs of, the procefs.

The

The Method of impregnating Water in large
Quantities *with Fixed Air, fo as to give it
the Properties of Mineral Water, for the
Ufe of the Sick on Board of Ships, and in
Hofpitals.*

DR. PRIESTLEY, fome years ago, com-
municated to the Lords of the Admiralty a
method of impregnating water with fixed air,
obtained from an effervefcing mixture of chalk
and vitriolic acid, and of making an artificial
Pyrmont water. This operation has fince
been confiderably facilitated by the invention
of Dr. Nooth's glafs machine, with Mr.
Parker's and Mr. Magellan's improvements.

That machine, though admirably contrived
for the preparation of fuch quantities of artificial mineral water as may be neceffary in
private families, would be too fmall for the
fickly crew of a large fhip. But it appears
to me that a mode may be adopted by which
the procefs may be performed on a much
larger fcale.

The

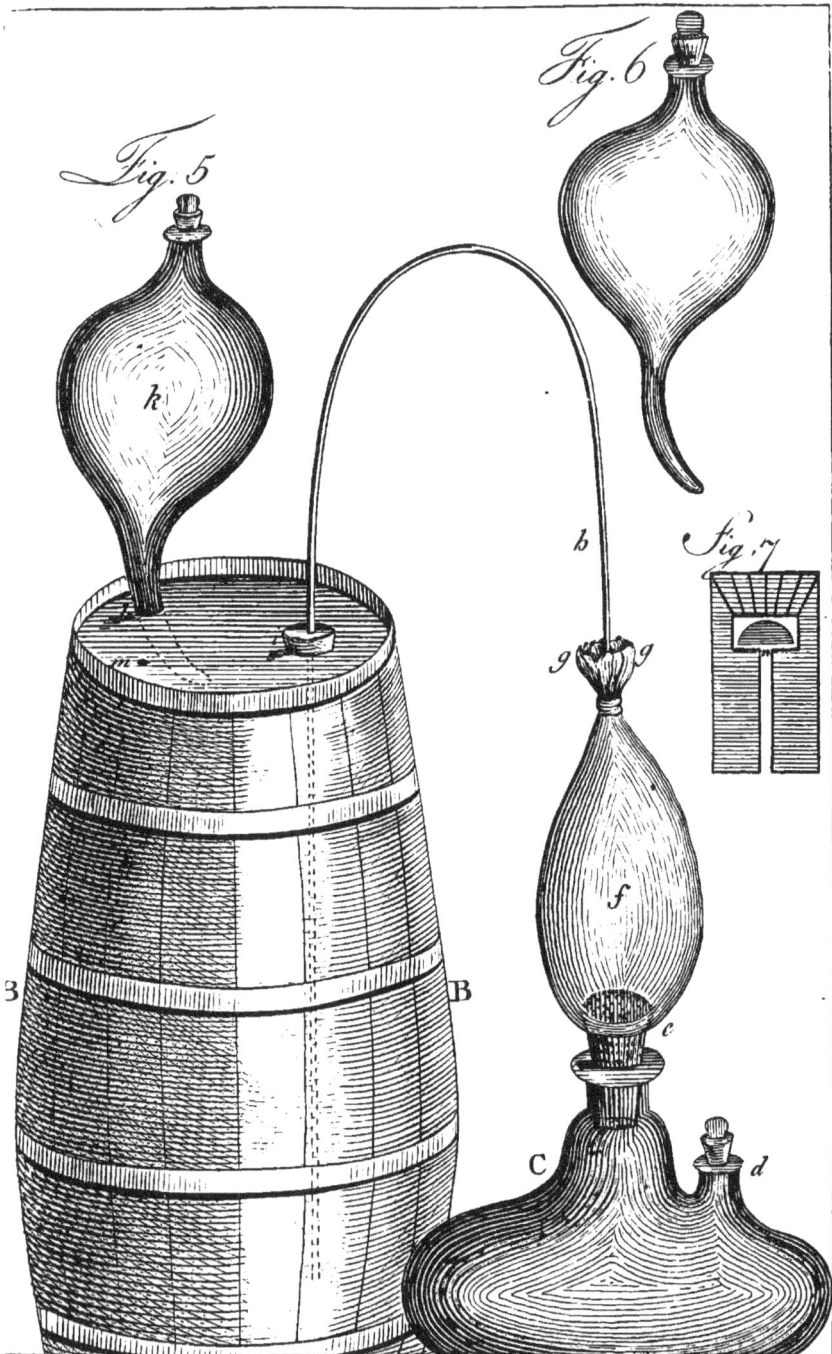

Fig. 5

Fig. 6

Fig. 7

k

m

b

g g

f

e

C

d

3 B

H. Cor.

Battenley sc.

The advantages which would proceed from an eafily practicable method of fupplying the fick men in long voyages with fuch water, muft be obvious to every medical practitioner. The mineral waters of Pyrmont and Seltzer may, by thefe means, be clofely imitated, and the artificial water will be beneficial in all cafes in which the natural is found ufeful. By this procefs alfo may Mr. Bewley's mephitic julep be prepared; than which the materia medica, perhaps, does not afford a more efficacious or more grateful medicine in putrid fevers, fcurvy, dyfentery, bilious vomitings, hectic, &c.

THE PROCESS.

Cut off the two extremities of a calf's or pig's bladder *(f)* (fig. 5.) and having previoufly moiftened them, into one end infert the top of the tubular ftopper *(e)* round the neck of which it is to be clofely faftened with ftrong thread. Into the upper end introduce the part *(g)* of the long bent tube *(h)* and tie them round in the fame manner. The pipe *(h)* muft be paffed through a hole,

C 2   formed

formed by a hot iron borer, in a large cork adapted to the orifice *(i)* in the cafk (BB) to which it muft be cemented: and the length of the pipe from this point muft be fuch as to reach within a few inches of the bottom of the cafk (BB) which is to be completely filled with frefh water, or fuch as has been recovered from lime.

To a quantity of mild calcareous earth and water, as directed in the preceding procefs, placed in the air veffel (C, fig. 5.) add a fmall portion of ftrong vitriolic acid, and by the time moft of the common air may be fuppofed to be expelled by the fixed air, arifing from the mild calcareous earth, add a larger quantity of acid, and putting the tubulated ftopper *(e)* in its place, the bladder *(f)* will become inflated. Prefs it gently till its fides collapfe; and then introducing the pipe *(bb)* with its cork, into the orifice *(i)* of the cafk (BB;) again prefs the air forward, as it diftends the bladder into the water cafk, where bubbling up through the water, it will rife to the furface, and by its preffure, force the water to afcend into the

funnel

funnel (k) which is to be cemented into the head of the cafk at (l). In proportion as the water in the cafk becomes impregnated with fixed air, that in the funnel will return into its place; but if, at any time, the latter fhould rife fo high as to be in danger of overflowing, a quantity of air may be let out of the water cafk, by means of the fmall plug at (m.) And this is neceffary to be done, ocafionally, to difcharge the refiduum of the fixed air, which is not foluble in water.

The water may be tafted from time to time, by drawing off a fmall quantity at a cock fixed into the cafk, and when it has obtained a fufficiently pungent tafte, the procefs may be finifhed. This will take feveral hours, but in this cafe little attendance will be re-quifite.* If the operation be required to be performed more expeditioufly, it may be quickened by agitating the water cafk. To do this, the tubular ftopper (e) muft be with-

* The operator muft be attentive that the top of the cafk be air tight. If fome water be poured upon it, any defects may be de-tected by the air bubbling through the water, and the faulty place muft be fecured with luting.

drawn

drawn from the air veffel, and fupported, together with the bladder, by an affiftant,
while the cafk (BB) is fhaken. During this
time another tubular ftopper muft be put into
the air veffel, and it may be immerfed into
a quantity of lime-water to prevent wafte.
When the agitation has been continued for
fome minutes, in proportion to the falling of
the water in the funnel, replace the ftopper
attached to the bladder (f) in the air veffel
when taken out of the lime-water, and proceed as before, repeating the agitation occafionally.

During the procefs, additional quantities
of vitriolic acid may be introduced into the
air veffel through the opening at (d) which
is to be, at all other times, carefully fecured
with its ftopper.

Perhaps the moft convenient fize for the
cafk, intended for the purpofe of impregnating water with fixed air, would be about
ten or twelve gallons. Should the fcurvy,
or other putrid difeafes, prevail; or fhould
putrid provifions or other feptic caufes render

der the crews more than ufually liable to
fuch difeafes, and occafion a larger confumption of this water to be neceffary, the cafk
may be proportionably larger, or a greater
number of fmall cafks may be employed.

I FLATTER myfelf that I have now
pointed out methods, not only of fupplying
the crews of his Majefty's fhips and others, in
every climate, with frefh water; but alfo of
affording them a medicinal water, which will
not only be a preventative againft putrid difeafes, but even a powerful remedy when they
actually exift.

The expence of precipitating the lime from
the water will be very trifling. I imagine
that, in performing the procefs in a large way,
eight ounces of mild calcareous earth, and
fix ounces of ftrong vitriolic acid will be fufficient for fixty gallons of lime-water. The
value of the firft is beneath notice, and the
prefent average price of oil of vitriol, at a
time that fulphur is very dear, is only five-
pence per pound. It is alfo to be confidered,
that the whole ftock of water will not need

to

to have lime added to it; that part only which is defigned for long keeping will require this treatment.

I would recommend that thefe operations be performed under the infpection of the furgeons of fhips and their mates; they may be conducted with facility; a little practice will make thofe gentlemen perfect mafters of the procefses; and, I am perfuaded, the liberal fpirit which prevails among the profefsors of the medical art, will not only prevent the rejection of an improvement, merely becaufe it is an innovation, but will incite them to reconcile difficulties, if it be pofsible they can occur, to promote the practice of what promifes fo much utility, comfort and falubrity to a body of men, on whom the enjoyment of our liberty, our property, and our religion fo eminently depend.

The malt decoction has been found, by experience, to anfwer in a great degree the beneficial intentions of the late truly amiable, humane and ingenious Dr. Macbride, who firft propofed the trial of it. And this falutary beverage may have its efficacy ftill farther improved, by impregnating it, in the above manner, with fixed air. Nay, I even believe that the decoction fo impregnated, and in-clofed in veffels for a few days, would ferment, and furnifh an ufeful and not unpleafant kind of beer.

I have repeatedly prepared an artificial yeaft, by impregnating flour and water with fixed air, with which I have made very good bread, without the affiftance of any other ferment. As I apprehend it to be a defideratum to procure frefh fermented bread at fea, at leaft that it would be an agreeable acquifition to the officers, the procefs, by which bread has been thus made, fhall be fubjoined. And if the above application of chemical facts, which were already known; if an improvement and extenfion of modes which have been already practifed, on a lefs enlarged plan, may, in any degree, tend to the prefervation of the health and lives of a part of mankind, more particularly expofed to difeafe, I fhall reflect with pleafure, while my capacity for reflection remains, that I have not lived unprofitably,

D

but

but have contributed my mite to promote the effential interefts of my country, and of humanity.

The Procefs for making artificial Yeaft.

BOIL flour and water together to the confiftence of treacle. When the mixture is become cold, fill a fmall cafk with it. This cafk is to be fitted up in the fame manner as that defcribed (BB, fig. 5.) for the impregnation of water with fixed air, and the procefs is to be conducted in a fimilar way, except that the cafk is to be agitated, as often as the mixture rifes to about two thirds of the capacity of the funnel (k); and after each agitation, which fhould continue during feveral minutes, the unabforbed air is to be let out, by withdrawing the plug from the orifice (m) till that part of the mixture which remained in the funnel have returned into the cafk. The orifice at (i) fhould alfo be larger than is neceffary in the former operation, on account of the fuperior vifcidity of the mixture. When, on the agitation being frequently

ly repeated, the mixture, which has afcended into the funnel, does not fubfide into the cafk, it may be fuppofed to be incapable of abforbing more air.

Pour the mixture, thus faturated, into one, or more, large bottle or narrow-mouthed jar. Cover it over loofely with paper, and upon that a flate or board, with a weight to keep it fteady. Place the veffel in a fituation where the thermometer will ftand from 70 to 80 deg. and ftir up the mixture two or three times in twenty-four hours. In about two days, fuch a degree of fermentation will have taken place, as to give the mixture the appearance of yeaft.

With this yeaft, when it appears to be in the above defcribed ftate, and before it have acquired a thoroughly vinous fmell, mix the quantity of flour you intend to make into bread, in the proportion of fix pounds of flour to a quart of the yeaft, and a fufficient portion of warm water. Knead them well together in a proper veffel, and covering it with a cloth, let the dough ftand for twelve hours,

or

or till it appear to be fufficiently fermented, in a degree of warmth equal to that above-mentioned. It is then to be formed into loaves, and baked.

Perhaps the yeaft would be more perfect, if a decoction of malt were ufed inftead of fimple water; but of this I have, as yet, had no experience.

The cafk in which the yeaft has been made, fhould be well wafhed as foon as the operation is finifhed, or it will contract a difagreeable taint.

The Procefs for making artificial Pyrmont Water.

TO every gallon of fpring water add one fcruple of magnefia alba, thirty grains of Epfom falt, ten grains of common falt, and a few pieces of iron wire, or iron filings. The operation is then to proceed as in the procefs for impregnating water with fixed air; and the water, if intended for keeping, muft be put into bottles clofely corked and fealed.

The

The Procefs to make artificial Seltzer Water.

ADD one fcruple of magnefia alba, fix
fcruples of foffil alkali, and four fcruples of
common falt to each gallon of water, and
faturate the water, as above, with fixed air.

To prepare Mr. Bewley's Julep.

DISSOLVE three drams of foffil alkali in
each quart of water, and throw in ftreams of
fixed air, till the alkaline tafte be deftroyed,
and the water have acquired an agreeable
pungency. This Julep fhould not be prepared in too large quantities; and fhould be
kept in bottles very clofely corked and fealed.
Four ounces of it may be taken at a time,
drinking a draught of lemonade, or water
acidulated with vinegar, or weak fpirit of
vitriol, by which means the fixed air will be
extricated in the ftomach.

REFER-

REFERENCES TO THE PLATES.

 F IG. 1. A A is the cafk in which the lime-
water is to be purified. It is reprefented as
tranfparent for the purpofe of fhewing the
fituation of the whole apparatus.

(a) Is the moveable top.

(b) The handle.

(c) A fmall hole to be exactly fitted with
a plug.

(dd) The ftrings by which the air veffel is
to be let down.

(ee) Two pegs, placed on oppofite fides of
the cafk, to which the ftrings are to be faftened before the cafk be ftopped.

(f) A cock to draw off the water.

Fig. 2. DD. The air veffel, fimilar to the
bottom part of Dr. Nooth's glafs machine.

(cc) A glafs ftopper, ground in to fit the
mouth of the veffel, having a number of
capillary tubes running from bottom to top
in a diverging direction, fo as to fpread the

air

air in its paffage through the water. Perhaps this veffel might be made· of pewter inftead of glafs, but in that cafe, as the tubes will be larger, a valve would be neceffary in the ftopper. I apprehend, likewife, that it may be formed of ftone-ware; but whatever be the materials, the veffel fhould be ftrong, and fo heavy as to fink readily in water, by its own weight.

Fig. 3. The ftopper viewed feparately to fhew its capillary tubes.

Fig. 4. A lead weight with an aperture in the middle, to flip over the ring of the handle *(b)* fig. 1.

Fig. 5. BB. A cafk whofe fides and ends muft be perfectly air tight, except two holes to be bored in the top.

C. The air veffel, the fame as fig. 2. but having an aperture at *(d)* which is to be fitted with a glafs ftopper, through which additional quantities of vitriolic acid may be introduced.

(e) The perforated ftopper, which muft have feveral circular or fpiral grooves round its

upper

upper part, to facilitate the faftening of the bladder *(f)* to it, the other end of which muft be attached in a fimilar manner to the broad end of the pipe *(h)* at *(gg.)*

(h). A long bent pipe made of pewter, having its lower extremity formed like the larger end of a clifter pipe. The length from the arch to the other extremity muft be fuch as to reach, nearly, to the bottom of the cafk BB. This pipe muft be cemented into a cork to fit the orifice in the cafk at *(i.)*

(k) The funnel, with a grooved cork or ftopper adapted to it, which is to be inferted and luted into the top of the cafk at *(l.)*

(m) The fmall hole, with a plug, to let out the portion of air, from time to time, which is incapable of being abforbed by the water.

Fig. 6. A feparate view of the funnel, which may be fabricated of tin, or copper covered with tin.

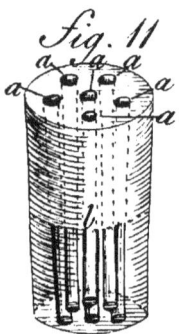

Fig. 11

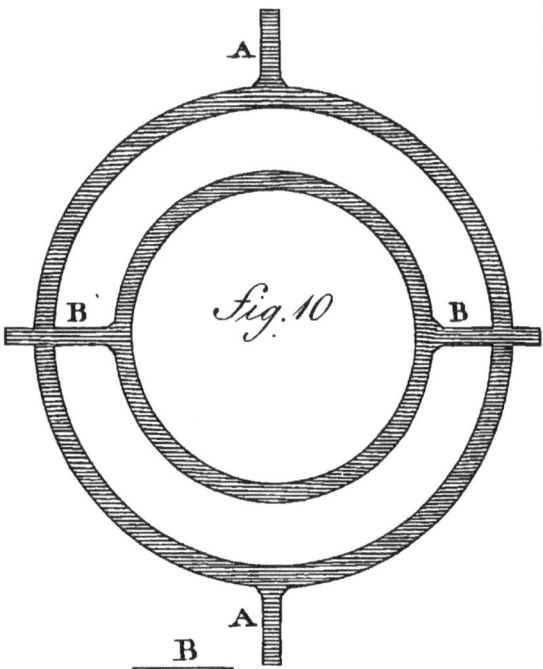

Fig. 10

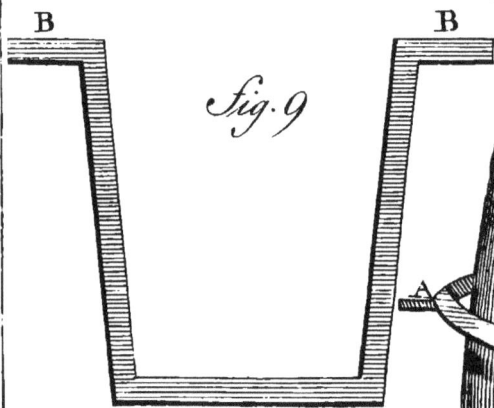

Fig. 9

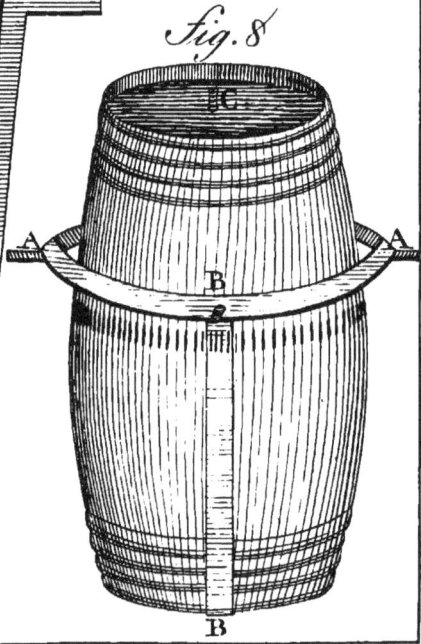

Fig. 8

Fig. 7. Is a longitudinal section of an earthen stopper, with a valve and capillary tubes communicating with a larger tube arising from its bottom. I am informed by the glafs-makers, that *diverging* capillary tubes are not easily formed in large bodies of glass; and I am induced to think that their diverging is not of so much importance as I at first apprehended.

The following figures are defcriptive of the mode of fufpending the cafk, which will be taken notice of in page 37.

Fig. 8. Is an elevation of the cafk, with its gimbols A A — and B, gimbol of the bearing hoop. C the bung — D the bearing hoop paffing under the cafk.

Fig. 9. The bearing hoop and gimbols B B.

Fig. 10. A plan of the bearing hoop and outer hoop, with the gimbols $\left\{ \begin{array}{c} B\,B \\ \text{and} \\ A\,A. \end{array} \right\}$

The fcale is an inch to the foot, fuppofing the cafk made ufe of to be a puncheon.

E   N. B.

N. B. The gimbols A A, muft be inferted into two ftanchions under any of the
hatchways of the fhip, and the infertion muft
be fo high, as to prevent the bottom of the
cafk from touching the deck below, by the
motion of the fhip, and thefe ftanchions muft
ftand tranfverfely or athwart fhips.

Fig. 11. Is a pewter ftopper with compartments *a a a a a*, into which fmaller glafs
ftoppers containing capillary tubes are to be
filled, and thefe are to communicate at *b*,
with paffages proceeding from the lower furface of the ftopper.

POST-

POSTSCRIPT.

AN unavoidable delay, in the publication of the preceding sheets, has afforded me an opportunity of submitting them to the perusal of several gentlemen distinguished for their philosophical knowledge, and to others who to science have had opportunities of adding long experience in maritime affairs. I am happy in having received from many a full, and from all a general approbation of my scheme; and hope it is now in my power to remove the few objections which have been made to it.

My particular acknowledgments are due to Dr. Lind of Haslar hospital, who was so kind to give the pamphlet a candid and attentive perusal. In a letter with which he has been pleased to favour me, he declares his opinion

that

that the method I have propofed for preferving water, at fea, from putrefaction, *well
deferves a public trial.* " I make no doubt,"
fays this ingenious phyfician, " but very
confiderable advantages would be derived
from it in fome fituations, but am afraid in
others, the agitation of the fhip would ob-
ftruct the procefs." This is the only objection which Dr. Lind makes *from himfelf*; he
then proceeds to ftate fuch as may perhaps
be made by *others.* " With fome people
alfo," he obferves, " who are accuftomed to
object to every innovation that is attended
with any trouble, and whofe principles they
do not comprehend, difficulties may be ftarted
about the prudence of impregnating all the
water with lime, and depending on a procefs
to render it fit for ufe, which tempeftuous
weather, the brittle materials of the veffels
employed, and even, fometimes, the neceffary duties of the fhip might interrupt for a
length of time. Their prejudices would even
urge the time required for purefying the
water, and the room taken up by an additional cafk or two, had for that purpofe upon
deck, as objections which, at leaft, it would

require

require time to remove. I take the liberty to point out the greatest objections that occur to me as being *capable* of being made against its use, that you may have an opportunity of satisfying *others* with respect to them, for with respect to *myself*, I think that all these inconveniences would be more than counter-balanced by having good water at sea, which must materially contribute to the health of the people who drink it."

I with pleasure embrace the opportunity which Dr. Lind has so kindly given me, of satisfying those in whose minds these doubts may have arisen, on perusing the proposed method. On mentioning the Doctor's apprehensions that the process might not be always practicable, in a rough sea, to captain William Roberton, formerly of the marines, a gentleman of great experience, and eminent for his skill in mechanics; actuated by principles of humanity and zeal for the service, he has, very obligingly, furnished me with a mode of suspending the purefying cask in such a manner, that it may always retain its level, whatever be the motion of the ship. A plan, &c.

of the apparatus to be ufed for this purpofe, from a drawing with which captain Roberton has favoured me, will be found in the third plate, to which, and the annexed defcription, I beg leave to refer the reader for farther information.

By this means the procefs may be carried on in all weather; but it is not neceffary it fhould be performed only *immediately when the water is wanted:* The operation may be continually going on, it will require very little attendance, and the water when freed from the lime will keep a confiderable time; much longer, I expect, for reafons, which will foon be adduced, than frefh water from the fpring.

Veffels made of ftone-ware, of fuch ftrength as to make them heavy enough to fink in water, would not be very liable to be broken. If the cautions I have given be properly ob-ferved, there will be no danger of explofion, even if the whole quantity of acid be *gradually* added at one time. It would be ingrati-tude in me to omit profeffing my obligations

to

to my very worthy and ingenious friend
Mr. Wedgwood, for his great and difin-
terefted attention in preparing models of the
air veffels, though he candidly acknowledges
the veffels may be more advantageoufly made
of what is commonly called ftone-ware. But
fhould thefe be found fubject to the inconvenience alluded to, every end will be anfwered by the ufe of pewter veffels.* The
principal difficulty with refpect to the em-
ploying of pewter in the fabrication of the
air veffels was the fuppofed impracticability
of forming capillary tubes in ftoppers of that
metal. Fig. 11. exhibits a view of one, in
which are feveral compartments to be fitted
with fmaller glafs ftoppers containing capillary
tubes; thefe communicating with other tubes
proceeding from the lower furface of the
ftopper.

A very ingenious correfpondent has fuggeft-
ed that inconvenience might arife, from the

* The beft London pewter contains but a very fmall portion of
lead, and confifts chiefly of tin or bifmuth; and as the affinity of
vitriolic acid to calcareous earth, fo far exceeds its affinity to lead;
there can be no danger of any faturnine impregnation to the
water.

E 4 precipitating

precipitating lime falling on the ftoppers and blocking up the capillary tubes. But it is to be confidered, that the conftant emiffion of air through thefe tubes muft prevent this impediment. In all the experiments I have made, fuch an event has never happened; but fhould it ever appear that this apprehenfion is well grounded, a piece of bladder perforated, in feveral places, with a fine pin or needle, or a double piece of fine linen cloth, tied tightly round the neck of the veffel, allowing the upper part to diftend into a convex form, will entirely preferve the tubes from being obftructed. Indeed, from feveral experiments I have lately made, either of thefe coverings, when tyed over a perforated cork, or pewter ftopper, or even over the mouth of the air veffel without either cork or ftopper, will ferve as a fubftitute for capillary tubes: and thus the danger of the procefs being impeded by accidents to the apparatus is almoft anni- hilated; for any large bottle, weighted fo as to fink in water, which may be done by a fufficient quantity of pebbles, and tyed over in the above method, may be employed as an air veffel for the precipitation of the lime.

I have

I have never been at fea; but fo little attendance does the procefs require that I cannot conceive it will ever interfere with the neceffary duties of the fhip. In the diftillation of fea water the conftant attendance of at leaft two men is required.—In the operation of precipitating the lime from the watèr, when the air veffel is immerfed, the procefs goes on of itfelf and during the whole time will fcarcely require fifteen minutes attendance. Thefe obfervations will alfo obviate the objection refpecting the length of time required for the operation. I believe I have in the preceding account extended it to the utmoft; but in a procefs which is almoft, felf-conducted, and continually going on, this objection can have no weight. As to the room taken up by the additional cafks, captain Roberton affures me they may be conveniently difpofed of under the hatchways and removed occafionally without any impediment to the bufinefs of the fhip. But if the advantage of being fupplied with fweet inftead of putrid water, be not fuperior to every trifling inconvenience that may be urged, it is not worthy the refearch.

I truft

I truft I have now anfwered every objection that can be made againft my propofals. One circumftance has appeared fince the printing of the former part of this account, which was entirely unexpected, but muft prove an additional recommendation of the method. Water is not only preferved from putrefaction, but meliorated from its original qualities by the operation, which has much the fame effect, in fome inftances, as diftillation upon it; with this advantage, that the water is not deprived of its air by the procefs, as it is when diftilled. Very hard water, from different pumps, having been impregnated with quick-lime, and afterwards freed from it by means of fixed air, was rendered foft as rain-water, afforded little or no precipitation when a folution of fixed alkali was added to it, and united readily and perfectly with foap: whereas, previous to the operation, it depofited a very large fediment on the addition of fixed alkali, and formed turpeth mineral on dropping in a fmall quantity of a folution of quick-filver in nitrous acid. But, notwithftanding thefe proofs that the water contained a vitriolic falt, it muft remain to be determined, by farther

ther experiments, whether the water, ufed in thefe trials, owed its hardnefs, *principally*, to fuch a falt, or, like the Rathbone-place water, to calcareous earth diffolved in it by means of fixed air. Lime-water mixed with it, in equal proportions, became turbid, and, when the precipitate had fubfided, the water was foftened.

THE END.

www.ingramcontent.com/pod-product-compliance
Lightning Source LLC
Chambersburg PA
CBHW031750090426
42739CB00008B/951